STARK LIBRARY SEP -- 2022

Gastornis

by Grace Hansen

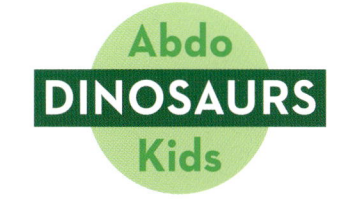

Abdo Kids Jumbo is an Imprint of Abdo Kids
abdobooks.com

abdobooks.com

Published by Abdo Kids, a division of ABDO, P.O. Box 398166, Minneapolis, Minnesota 55439. Copyright © 2021 by Abdo Consulting Group, Inc. International copyrights reserved in all countries. No part of this book may be reproduced in any form without written permission from the publisher. Abdo Kids Jumbo™ is a trademark and logo of Abdo Kids.

Printed in the United States of America, North Mankato, Minnesota.

052020
092020

THIS BOOK CONTAINS RECYCLED MATERIALS

Photo Credits: Alamy, iStock, Science Source, Shutterstock, Thinkstock, ©Vince Smith p.21 / CC BY-SA 2.0

Production Contributors: Teddy Borth, Jennie Forsberg, Grace Hansen
Design Contributors: Dorothy Toth, Pakou Moua

Library of Congress Control Number: 2019956465

Publisher's Cataloging-in-Publication Data

Names: Hansen, Grace, author.

Title: Gastornis / by Grace Hansen

Description: Minneapolis, Minnesota : Abdo Kids, 2021 | Series: Dinosaurs | Includes online resources and index.

Identifiers: ISBN 9781098202439 (lib. bdg.) | ISBN 9781098203412 (ebook) | ISBN 9781098203900 (Read-to-Me ebook)

Subjects: LCSH: Gastornis--Juvenile literature. | Dinosaurs--Juvenile literature. | Flightless birds--Juvenile literature. | Paleontology--Cenozoic--Juvenile literature. | Dinosaurs--Behavior--Juvenile literature.

Classification: DDC 567.90--dc23

Table of Contents

Gastornis 4

Habitat 6

Body 8

Food 14

Fossils 18

More Facts 22

Glossary 23

Index 24

Abdo Kids Code 24

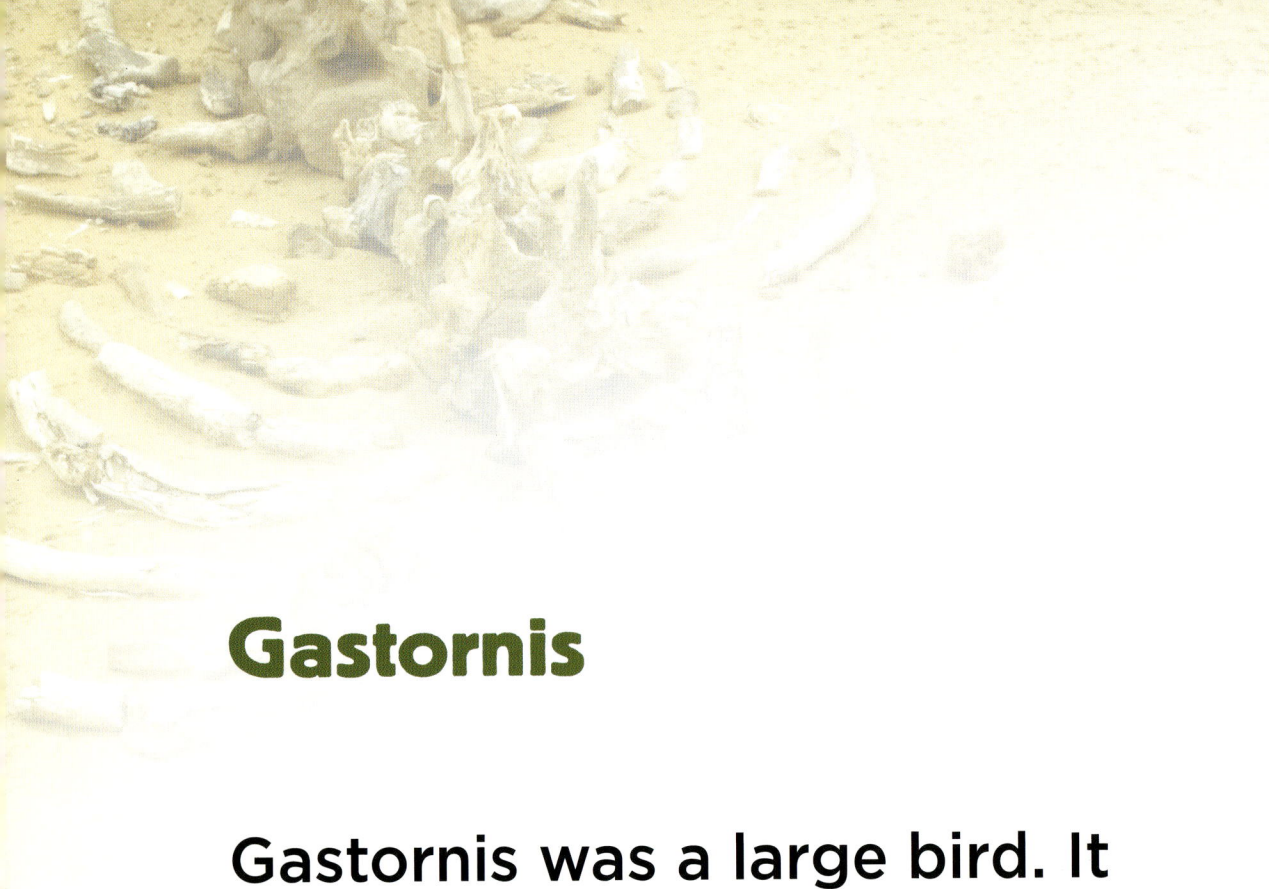

Gastornis

Gastornis was a large bird. It lived around 56 million years ago during the Cenozoic era.

5

Habitat

Gastornis lived in tropical areas. It could be found roaming rain forests.

7

Body

Gastornis was a huge bird! It stood at least 6 feet (1.8 m) tall on two thick legs.

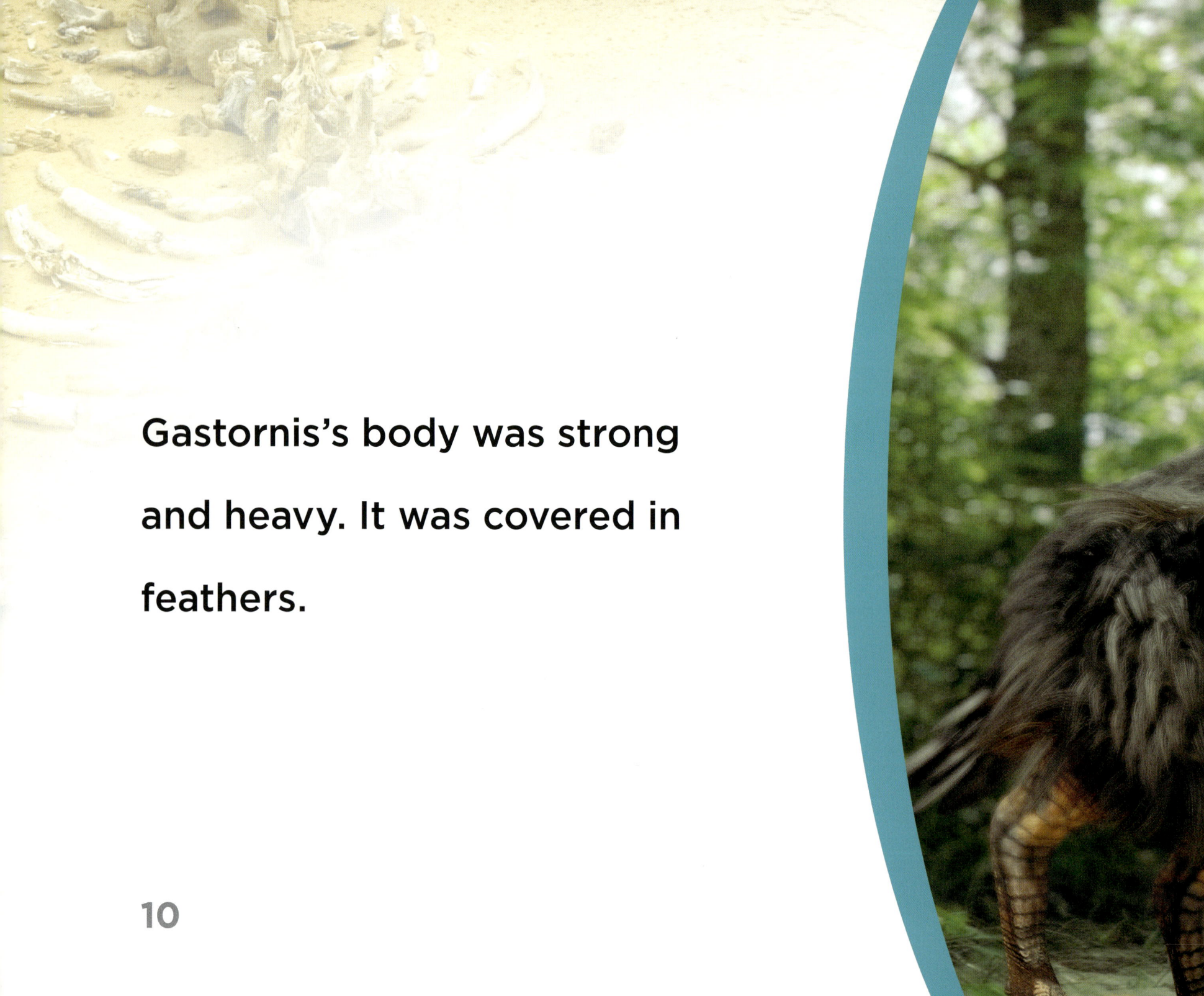

Gastornis's body was strong and heavy. It was covered in feathers.

11

Even though Gastornis was a bird, it could not fly. Its tiny wings were useless.

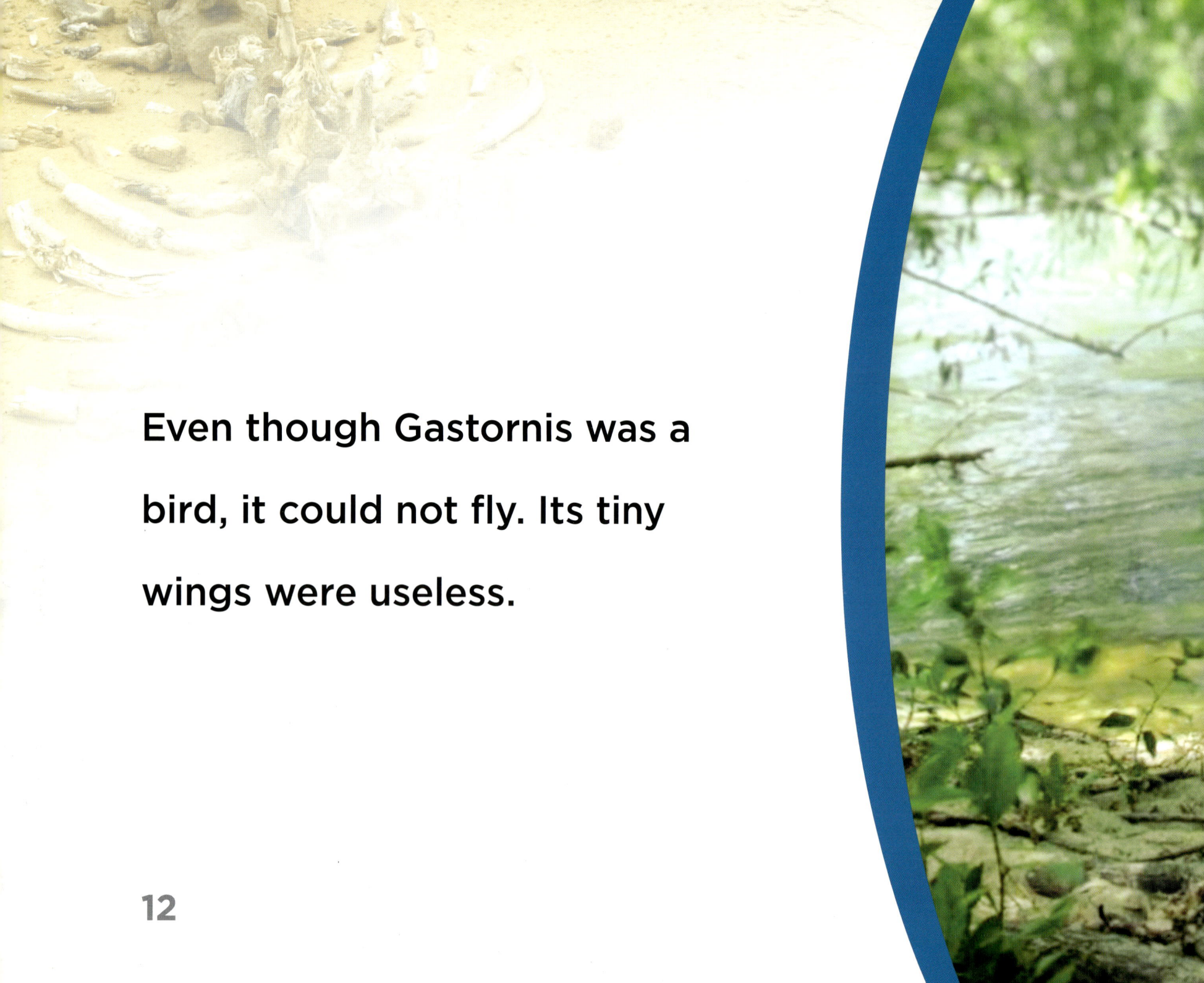

13

Food

Gastornis had a large, powerful **beak**. It used its beak to break shelled fruits and seeds.

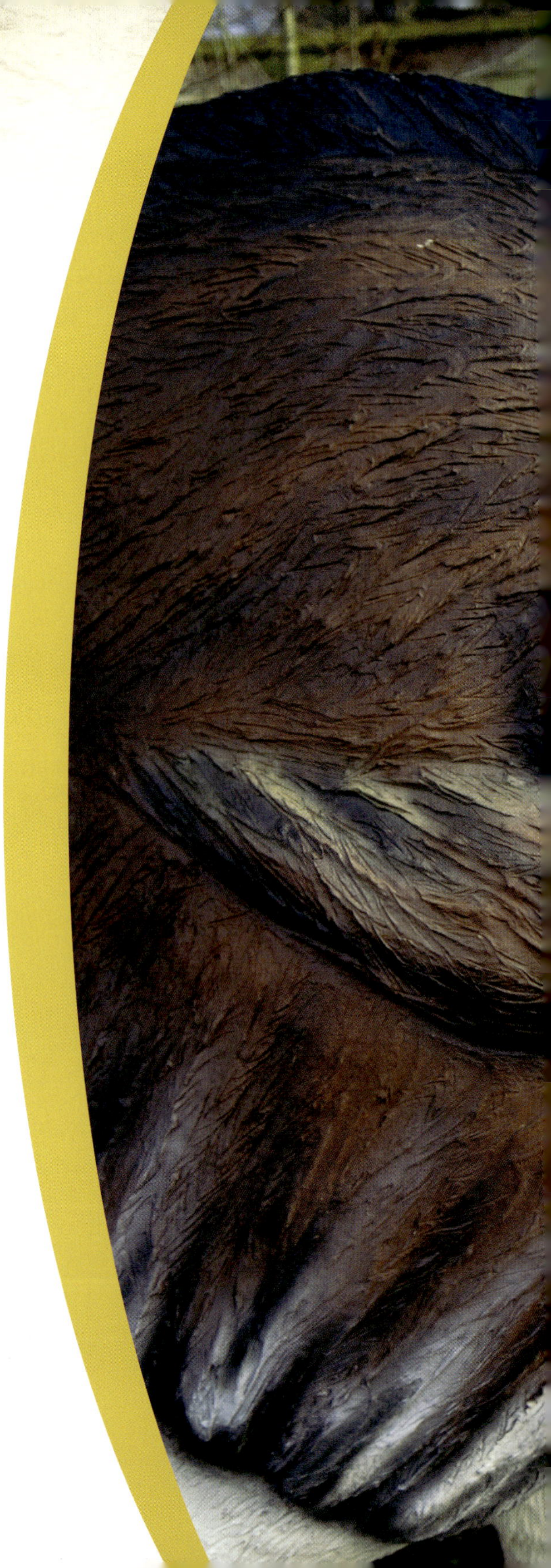

Scientists once thought it was a meat-eater. But they have now learned the bird was a **herbivore**.

Fossils

Gaston Planté found the first Gastornis fossils. He was a French scientist. In 1855, the prehistoric bird was named after its discoverer. Its name means "Gaston's bird."

19

Many egg **fossils** were found in France. Gastornis eggs were a bit larger than an ostrich's eggs.

More Facts

- Gastornis likely weighed around 400 pounds (181 kg).

- Scientists learned about what Gastornis ate from its remains. Its remains showed no evidence that the large bird ate meat.

- Another clue that Gastornis was a herbivore came from its beak. Though it was large and powerful, it did not have a hook at the end. Beaks made for meat-eating usually have a hooked tip.

Glossary

beak – the hard-pointed part of some dinosaurs' mouths.

Cenozoic Era – the third of the major eras of Earth's history, beginning about 65 million years ago extending to the present. It began after the mass extinction event that ended the Cretaceous period.

fossil – the remains, impression, or trace of something that lived long ago, as a skeleton, footprint, etc.

herbivore – an animal that only feeds on plants.

prehistoric – belonging to a period in a time before written history.

Index

beak 14

body 10

eggs 20

feathers 10

flight 12

food 14, 16

France 18, 20

habitat 6

height 8

legs 8

Planté, Gaston 18

size 4, 8, 10, 12, 14

wings 12

Visit **abdokids.com** to access crafts, games, videos, and more!

Use Abdo Kids code **DDK2439** or scan this QR code!

3 1333 05175 1996